Growing Magic Mushrooms

A New Indoor Growing Technique

Steven McKenna

Contents

Introduction.

Introduction

You don't need to grow mushrooms in an expensive sterile environment. When I first learnt to grow mushrooms, I was told that you needed a sterile environment to grow them reliably costing thousands.

My guide on how to set up a low-tech mushroom farm goes into more detail on this, but to summarize: there is a much easier way to grow mushrooms than is usually taught.

You can pasteurize straw with simple techniques.

It's sooo much cheaper and quicker to get setup and growing, and much easier to learn.

This is a huge part of the reason why we love teaching how to grow mushrooms like this – it just makes it so much more accessible for people to get started.

INDOOR MUSHROOM GROWING TECHNIQUE FOR THE BOIL - A - BAG.

Firstly: Getting some "Boil a Bags" is all important in using this technique to grow your mushrooms. Boil a bags are plastic bags that can go in the pressure cooker or microwave without melting. the brand i use are made by a company called Lakeland Plastics, in packs of 50 at a size of 8 x 12 or that's 20cm by 30cm: cost is about £2.50p per pack of 50.

Fig 1

Lakeland Plastics / Alexandra Buildings / Windermere / Cumbria / LA23 1BQ / UK
TEL: 015394 88100
Barcode number : 2001 4056

Here are some essential instruments you will need to grow your mushrooms

Top row left to right

1. a spore print in a jar \ 2. propyl alcohol \ 3. spirit burner \ 4. glass pyrex container

Bottom row left to right

1. some cotton wool buds \ 2. A flat edged metal scraper \ 3. metal tweezers \ 4. metal scissors \ 5. syringe needle \ 6. syringe

A long syringe needle can be easily made by removing the original needle from its green plastic holder with the use of a pair of grips & replacing with a thin piece of stainless steel tubing, obtainable from model and craft shops, or metal engineering companies. You may find it easier if you heat the tip of the steel tube before you insert it into the holder.

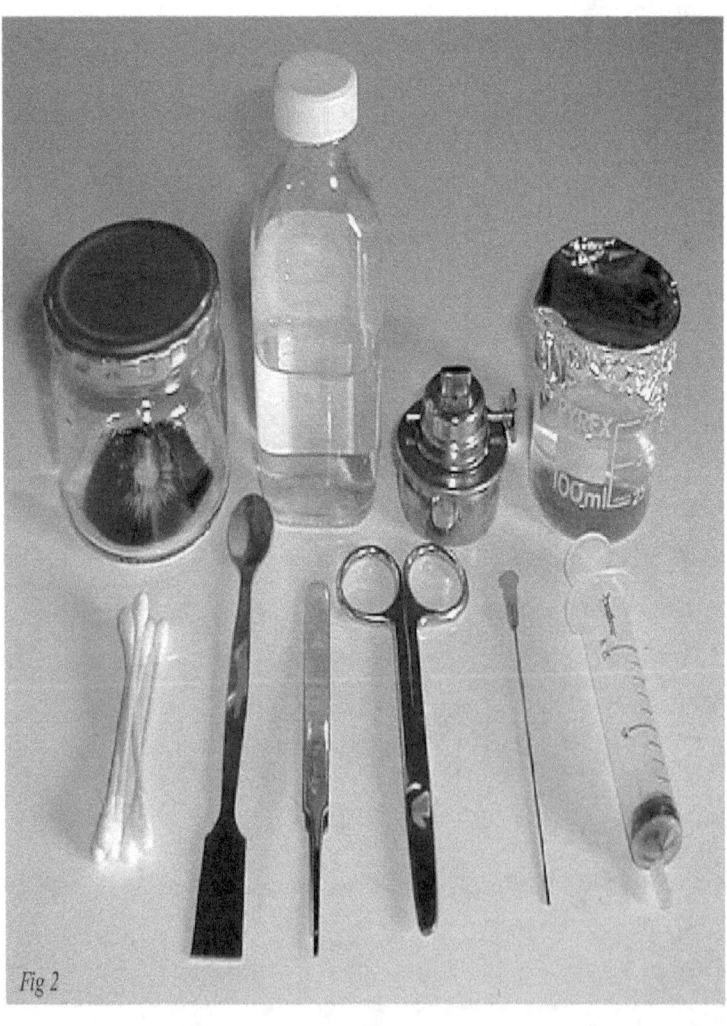

Fig 2

Part 1: Making The filter can

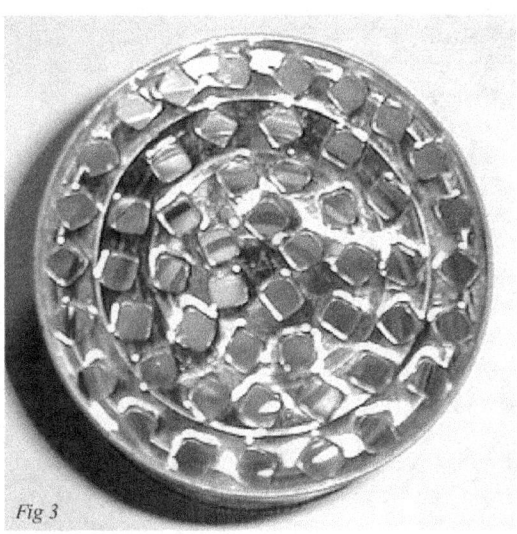

Fig 3

Fig 3. Get an empty cleaned 3 to 4 inch round food can *"when you first open the can, don't open all the way, leave the lid connected by an inch or so" this will serve as an attached lid".*
Now with a hammer and nail, make holes all over the bottom of the can, so that is has lots of breath holes.

Fig 4

Fig 4. Line the bottom inside of the can with about 1cm of rock wool, or ordinary loft insulation fiber glass,

Fig 5

Fig 5. Fill the can with Vermiculite untill you are 1cm from the top.

Alternitively, instead of using vermiculite, Simply fill the can with small pieces of rockwool.

Fig 6

Fig 6. Top with a little more rock wool to hold in place.
The filter is now completed & with the lid closed down is ready to be used.

Part 2: Prepareing the Straw

Fig 7

Fig 7. Chop some straw to be at 3 to 10cm lengths " you can buy straw pre cut to this length from many pet stores". Place as much of the straw as you need into a large water tight container.

Fig 8. A heaped table spoon of marmite is added to a pouring jug. *Marmite, vegimite, or yeast extract' is used to help quicken the germination time of the spores.*

Fig 9. The marmite is dissolved in some boiling water.
This is then poured into a bucket & an additional 6'ltrs of of boiling water is added.

Fig 10. The marmite mixed boiling water is poured over the straw in the yellow bucket, the straw will now be weighed down by filling the white bucket with water.

After an hour check to see how far the straw has been pushed down into the bucket & that the water level is still covering the straw. If needed add more boiling water & leave to soak for 1 more hour.

Fig 10

Fig 11 After the straw has been soaked, it is turned out in the bath tub & left for 1 hour to thoroughly drain.

Fig 11

Fig 12. Evenly spread out the straw in the bath tub, if you find the straw is still to wet, then use your hands to squeeze out the surplus water.

Fig 12

Fig 13. Some brown rice flour for extra food is sprinkled over the straw *"for 6 bags worth, Sprinkle around 3 to 4 handfuls"* then thoroughly mix in using your hands.

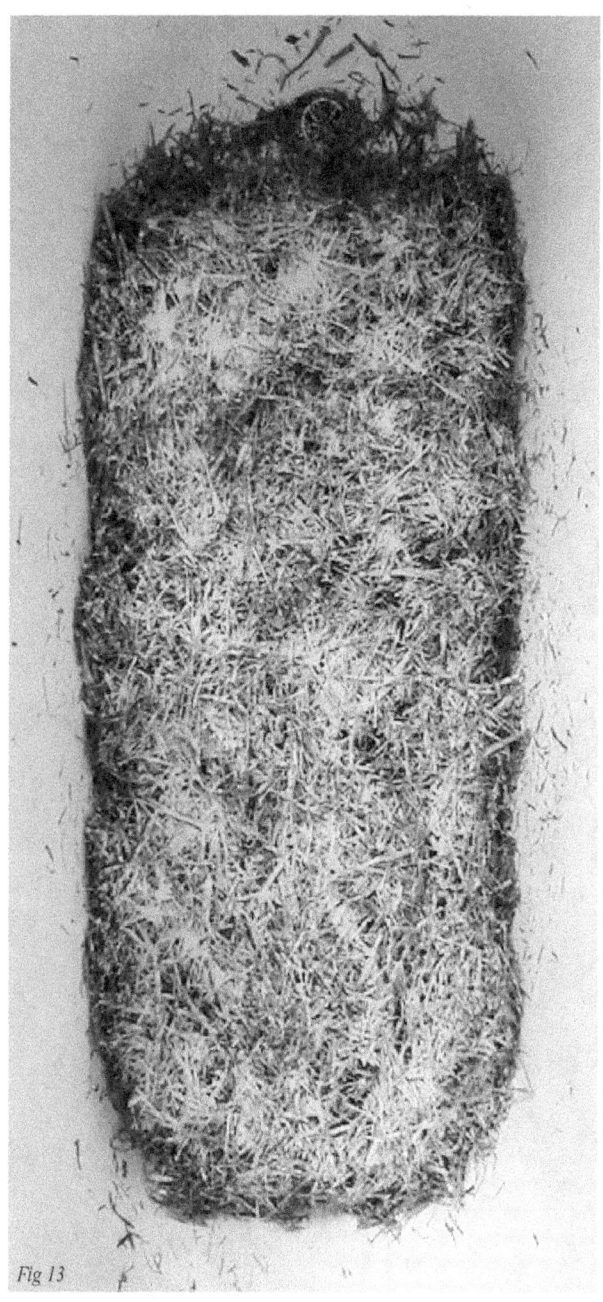

Fig 13

Part 3: Filling the bags

Fig 14

Fig 14. Firmly pack some straw mixture into the bags until they are almost full, leaving enough room to fit the filter can into the top of the bag.

Fig 15. The filter can is fitted into the top of the bag & secured by going around the neck of the bag 3 times with wide masking tape, then firmly push the tape onto the can & bag.

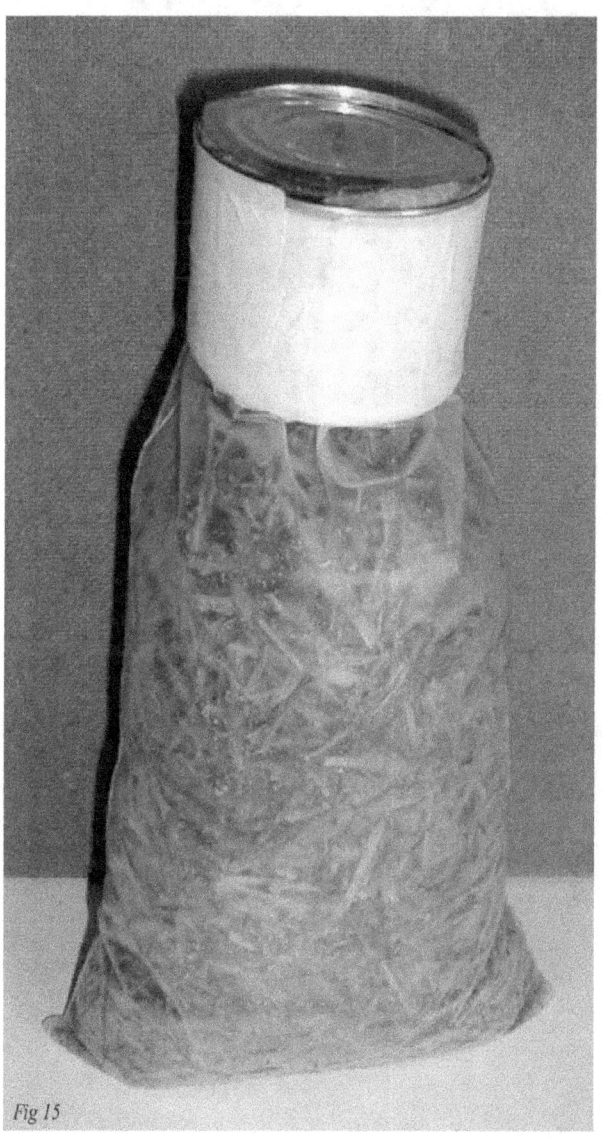

Fig 15

Part 4: Sterilizing the bags

Fig 16. All 6 bags are now ready to be sterilized.

Fig 17. Two bags are placed into the pressure cooker *"Maybe only one, depending on the size of your cooker"* water is added & with the lid on, the bags are brought up to maximum pressure "15psi" for 60 minutes, You May find it favorable to do this in two' 30 minute stages.

Part 5: Preparing bags for inoculation

If you are planning on using a standard 4cm long needle, then please follow this method for short needle inoculation.

Fig 18. These 2 bags have just been sterilized & have thoroughly cooled down. Tape 2 supports, "*1 each side of the bag*" from can to bag base' as illustrated on the bag to the left.

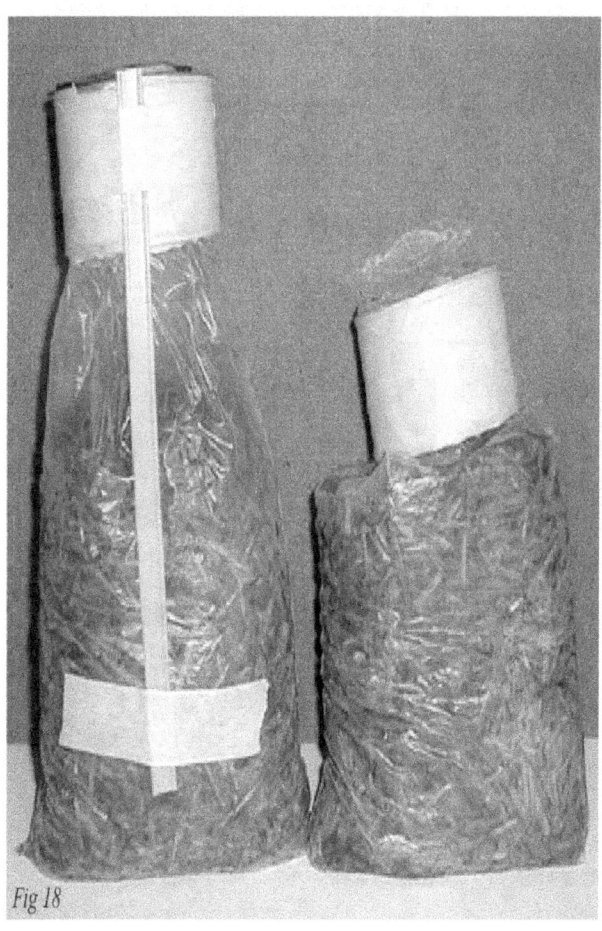

Fig 18

Fig 19. Here we see 4 bags that have had their supports added, it can be helpful to stand them in some kind of container, this will make things easier for you to prepare & inoculate them, it also makes sense at this point to add tape to the inoculation points, 1 on each side of the bag "2 inoculation Points per bag" *"this piece of tape stops the hole from enlarging upon inoculation"* an additional piece of tape is placed & looped back, so that when you have inoculated the bag, you can quickly roll the tape down over the hole as you remove the needle from the bag.

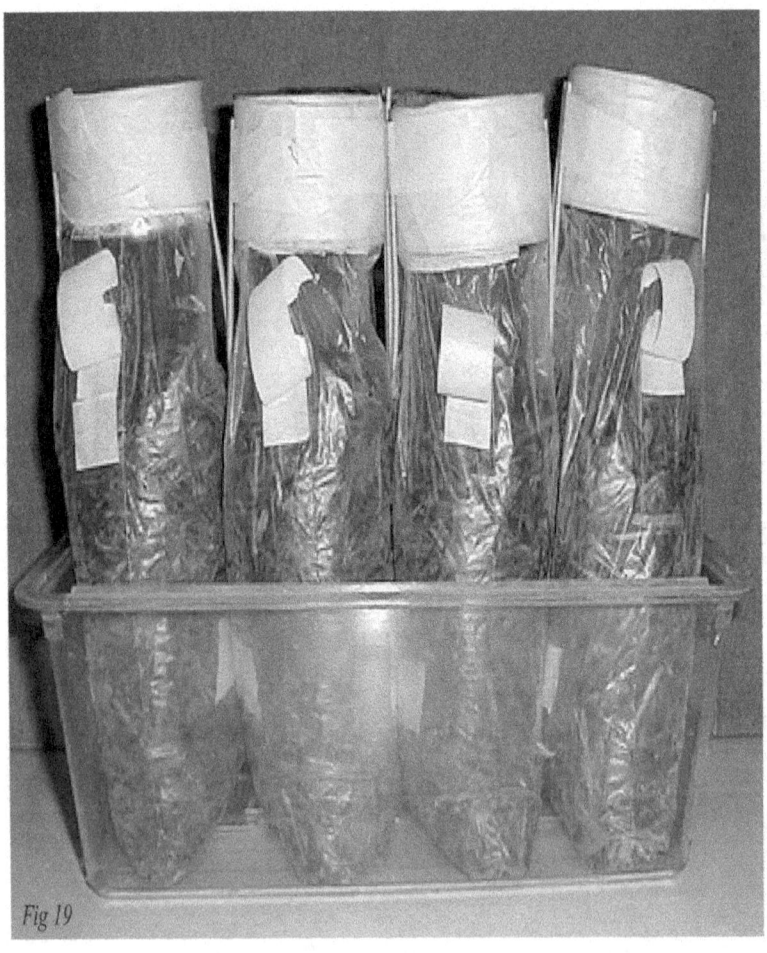

Fig 19

Part 6: Preparing the spore syringe

Fig 20

Fig 20. The syringe & its needle are connected & wrapped in tin foil, The pyrex jug in filled with 80ml of water & has a tin foil lid placed on top, The contents of the cooker is now brought up to full steam "15psi" for 20mins.

WARNING !
THE REST OF THE PROSESS WILL REQUIRE YOU TO WEAR A
PROTECTIVE BREATHING MASK, OR A CLEAN HANKICHIEF
COVERING YOUR NOSE & MOUTH.
THIS IS NOT FOR YOUR SAFETY, IT IS' SO THAT YOUR BREATH
DOES NOT DIRECTLY INFECT THE SPORE SOLLUTION

Part 7: Inoculating the bags
Short Needle Method

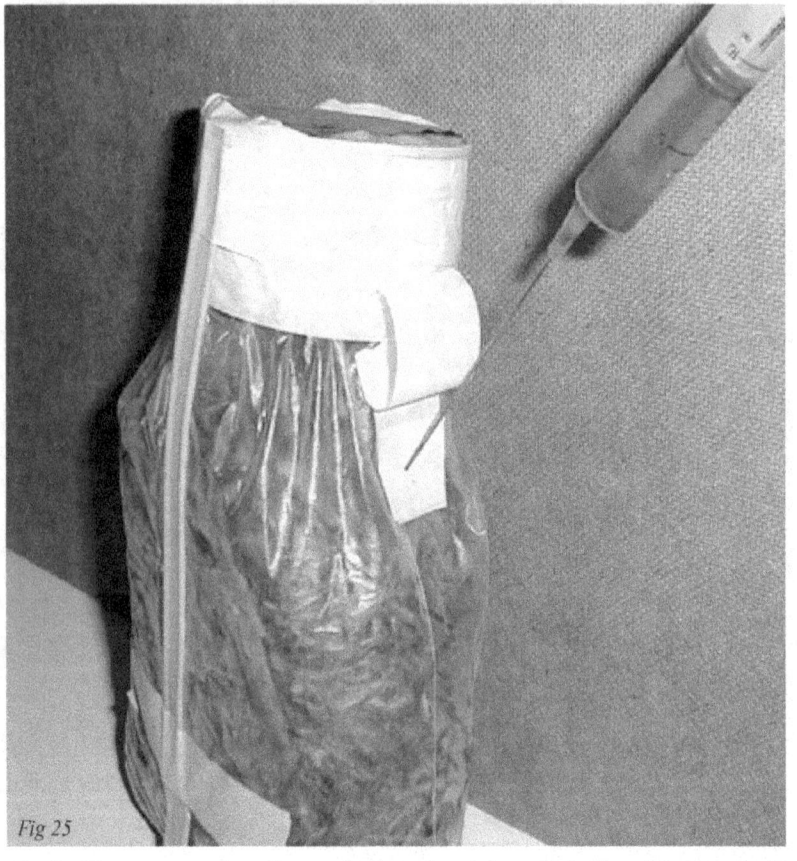

Fig 25

Fig 25. Here we see a bag about to be inoculated through the bag & bellow the filter level, this method is for short needles that cannot be passed all the way through the filter can for inoculation, there has also been a little preparation before this is done "see Fig 19".
Using a cotton wool bud swab both inoculation points with propyl alcohol, now that the needle is fully flame sterilized.
."Again heat the needle tip only! Until it is Red, this will allow you to easily penetrate the bag"
Insert the needle all the way into the bag & release 5cc of the spores into each inoculation area, swiftly rolling down the tape as you remove the needle from the bag.

Part 8: Inoculating the bags
Long Needle Method

This Method is much easier, quicker & more trouble free than the short needle method.

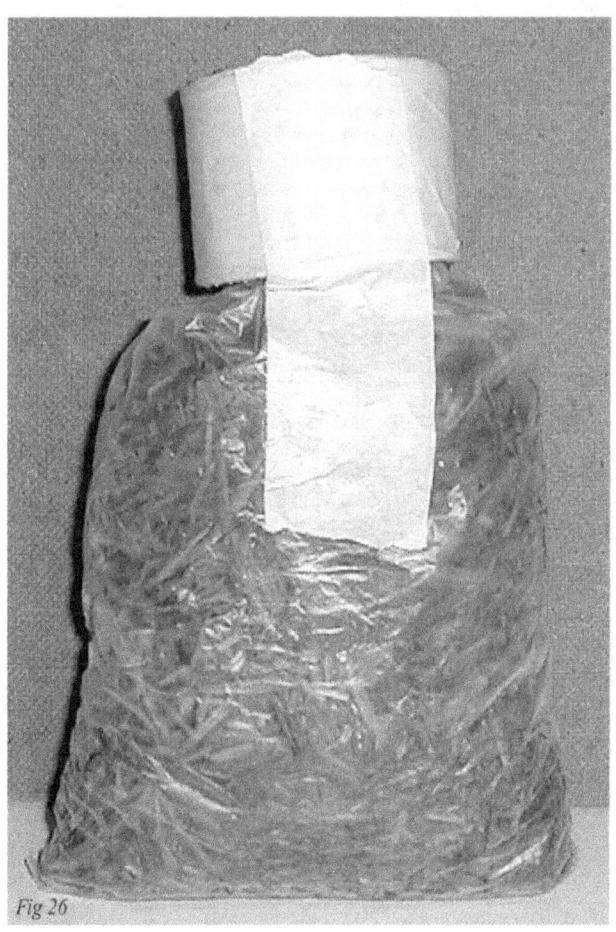

Fig 26

Fig 26. To secure it a sterilized & cooled bag has its filter can taped on both sides of the bag.

Fig 27. The lid of the filter can is lifted, the needle has been flame sterilized & is passed all the way through the can, & 10cc of spore solution is inoculated into the Straw, The needle is then removed & the lid of the can is closed down, if the lid keeps tending to lift up, then secure it with a small piece of tape.

The bags can now be placed in a cardboard box & kept in a warm place for spore germination & mycelium growth.

Part 9: Keeping an eye on things

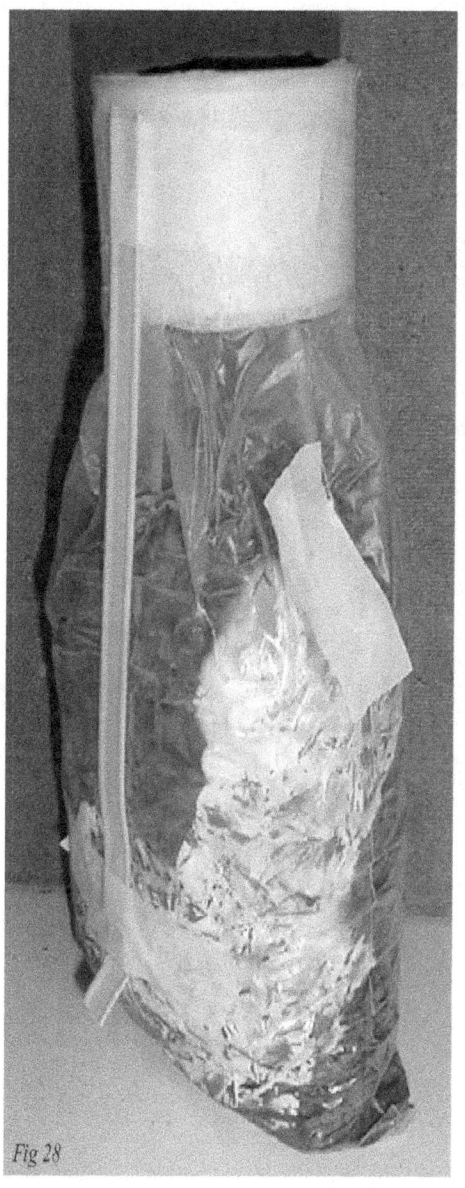

Fig 28.

Fig 28. A few days after inoculation the mycelium growth is easily visible.

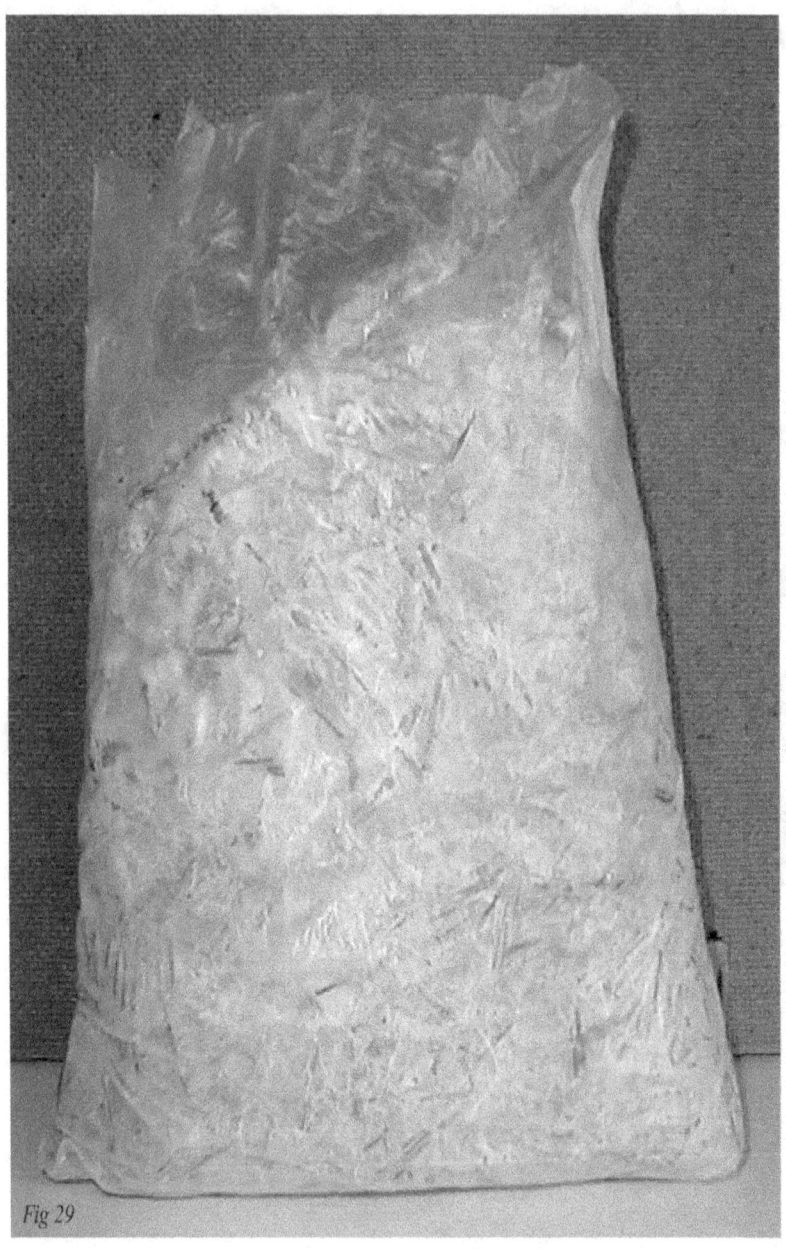

Fig 29

Fig 29. The Mycelium has totaly colinized the bag, It is time to remove the filter &
place them in the humidity chamber.

Fig 30. Place the bags into a humidity chamber, this is simply a 4ft by 2ft fish tank, also note that I have added a layer of soaked hydro pellets to the bottom of the tank, this will keep the interior nice & humid, there has also been a layer of plastic placed over the top of the tank to create a seal, so that none of the humidity escapes & no fly's can get in.

All that needs to be done for the next few days is to fan some fresh air into the tank twice a day & keep an eye out for signs of forming mushrooms.

Fig 31

Fig 31. Here we see the first signs of mushrooms starting to form, it is time to cut the bag down the sides & remove as much of the bag as needed, I like to leave at least 2cm of the bottom of the bag intact, this allows me to stand the bags anywhere I like, without the worry of them getting to wet when they are standing in the humidity chamber.

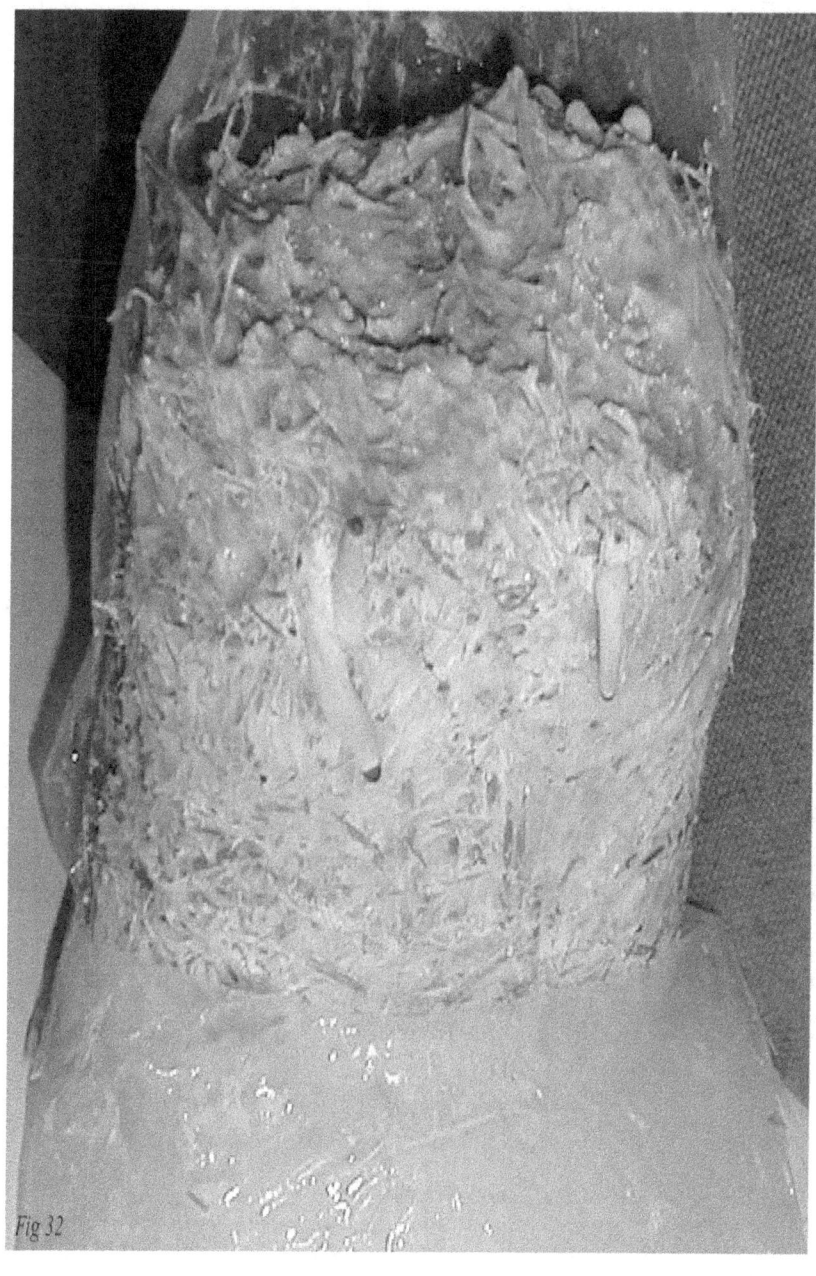

Fig 32. Here again we see the first signs of small mushrooms starting to form.

Part 10: Taking a sprore print

Fig 35

Fig 35. Now that you have grown some mushrooms, you will need to take some spore prints so that you can grow your next crops. First of all get some small clean glass jars with metal screw lids, take the lids off and lean them aside the jars & on a metal baking tray put them under the grill of your oven for 30mins at half the maximum heat so that they can be sterilized. When the jars have cooled, carefully go into the grill & place a lid loosely on each of the jars, now remove the tray from the grill & tighten the lids onto each of the jars.

Take your jars, tweezers, scissors & meths burner, to your mushroom growing tank !

FOR THIS PART OF SPORE PRINT TAKING WEAR YOUR BREATHING MASK

Get one of the jars and unscrew its lid *"but leave it on"*
Flame sterilize your tweezers & scissors. Gently push the tweezers into the top of an opened spore loaded cap & cut it from the stem with your scissors, now carefully place the cap gills down into the jar & immediately replace the lid ! but do not ! tighten it all the way, as any moisture in the jars or on the spores will need to naturaly evaporate. see Fig 37.

Fig 36. The cap has been carefully placed into the sterile jar & with the lid loose is left for at least 24 hours' for the spores to settle on the bottom of the jar.

Fig 37. In this picture, using the tweezers the caps have been removed from the jars & for a further 24 hours have stood in a clean draught free area, so that the spores can loose their moisture, now this has been achieved the lids of the jars can be firmly tightened for later use. **The Circle Is Now Complete !**

Fig 38 & 39. Here are some pictures to help you make an easy self humidifying chamber, most things to make the chamber can be obtained from local hardware shops & the plastic car registration plate nuts & bolts from car accessory shops. All that is needed to keep the tank humid is an air stone some tube & a small aquarium air pump, Simply make a small hole near the top of the tank & feed the pipe through to the water container, attach the air brick & pump. The inner angled drip lids are simply shaped 2mm plastic sheet.

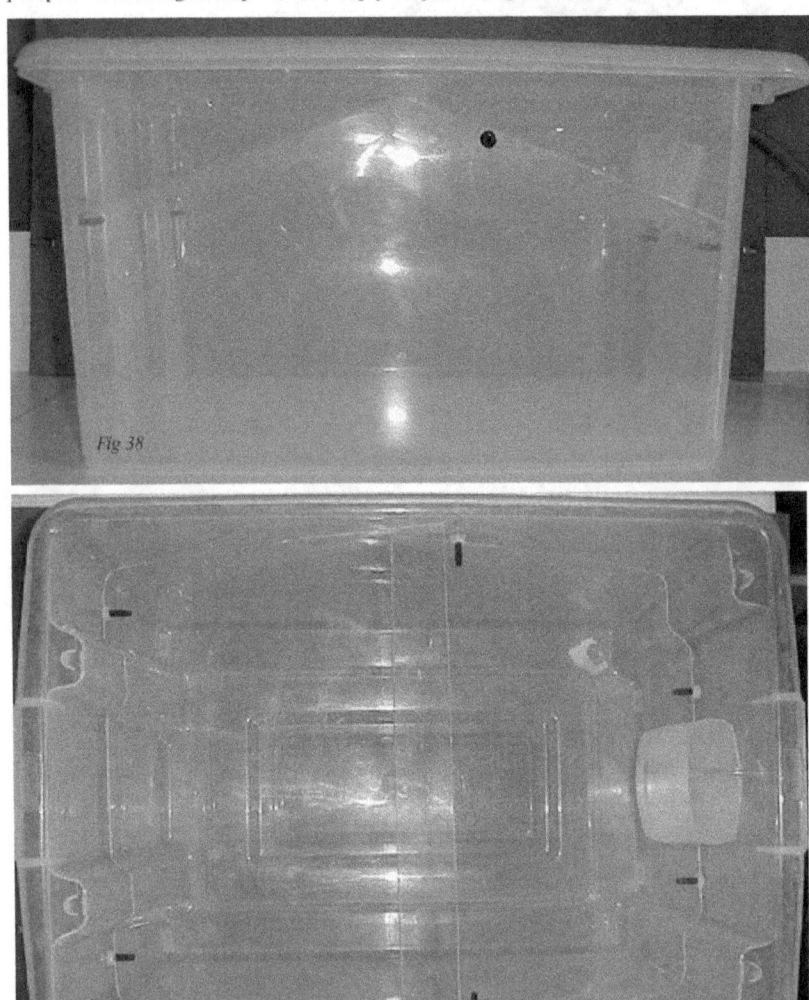

Fig 38

Fig 39

Fig 40 & 41. These two pictures show how to make a warm humid air delivery chamber, this is very useful for those who live in a cold climate, or want to grow at cold times of the year. All of the small connection ducts are potential outlets & the one large duct is the main input. The container is filled with a few Inches of water & with the lid & input fan fitted is ready to use, it may take you a few hours to set the water heaters timer to get the desired humidity temperature, when you have achieved this, plumb in the growing chambers & away you go !